MW00476766

A Guide to Collecting Wild Herbs

Written and Illustrated by
Julie Gomez

ISBN 0-88839-390-3
Copyright © 1996 Julie Gomez

Cataloging in Publication Data
Gomez, Julie, 1964-
   A guide to collecting wild herbs
   ISBN 0-88839-390-3

   1. Herbs—Identification. 2. Herbs—Collecting and preservation.
   I. Title.
   SB351.H5G65 1996   581.6'3   C96-910030-2

All rights reserved. No part of this publication may be reproduced,
stored in a retrieval system or transmitted, in any form or by any
means, electronic, mechanical, photocopying, recording, or otherwise,
without the prior written permission of Hancock House Publishers.
Printed in Canada

Production: Myron Shutty and Nancy Miller
Editing: Myron Shutty

Published simultaneously in Canada and the United States by

HANCOCK HOUSE PUBLISHERS LTD.
19313 Zero Avenue, Surrey, B.C. V4P 1M7
(604) 538-1114   Fax (604) 538-2262

HANCOCK HOUSE PUBLISHERS
1431 Harrison Avenue, Blaine, WA 98230-5005
(604) 538-1114   Fax (604) 538-2262

Contents

Glossary 4

Introduction 6

White Flowers

    Chickweed 8

    Field Peppergrass 10

    Ox-Eye Daisy 12

    Queen Anne's Lace 14

    Strawberry 16

    Watercress 18

    White Clover 20

    Yarrow 22

Blue, White or Pink Flowers

    Chicory 24

    Musk Mallow 26

    Nodding Wild Onion 28

Pink to Purple Flowers

    Bull Thistle 30

    Burdock 32

Pink to Red Flowers

    Fireweed 34

    Milkweed 36

    Red Clover 38

Yellow Flowers

    Dandelion 40

    Evening Primrose 42

    Mullein 44

    Purslane 46

    Scotch Broom 48

    Wild Lettuce 50

    Winter Cress 52

Green Flowers

    Curly Dock 54

    Lamb's Quarters 56

    Plantain 58

    Stinging Nettle 60

Brown Flowers

    Cattail 62

Glossary

Annual:	plants that complete their life cycle in one year: growing from seed, flower, set seed and die.
Axil:	the V-shaped angle between the stalk and the stem where they join.
Biennial:	plants that complete their life cycle in two years: developing leaves in their first year, producing stalk, flowers, set seed and dying the second year.
Bract:	small modified leaves near the flower.
Bulbet:	miniature bulbs that are clustered near the flowers.
Calyx:	fused or open sepals that support the flower.
Category:	plants are categorized by flower color: white, blue, pink, purple, red, yellow, green and brown.
Compound:	a division of leaves that form groups of three or more.
Disk:	the flat compact area in the center of a daisylike flower.
Edible parts:	the portions of the plant that are edible.
Habitat:	where to locate the plant described.
Harvest:	the season in which the plant should be harvested.
In bloom:	the season in which the flowers appear.
Linear:	leaves that are slender and simple.
Margin:	leaf edges.
Medicinal value:	the plant's healing capability.
Midrib:	the vein through the center of a leaf.

Node:	swollen joints on the stalk where shoots, buds and leaves form. (Roots usually form at the nodes on creeping stalks.)
Opposite:	leaves that develop at the same point on the stalk and stem.
Ovate:	leaves that have a broad base.
Perennial:	plants that continue their life cycle from new growth and live more than two years.
Pistil:	reproductive organ of the female flower.
Pollen:	male reproductive cells.
Rosette:	leaves that form on the ground before the stalk.
Seedpod:	a protective casing that contains the seeds before their release.
Sepals:	the (typically) green petals of the calyx that enclose the developing flower and that support the opened flower.
Species:	both the common and scientific names are given.
Stigma:	the sticky, coated pistil that accumulates pollen.
Umbel:	clusters of flowers that form an umbrella-like shape.
Use:	indicates cooking times and preparation.
Warning:	this is given when a plant resembles a poisonous species, causes dermatitis, or may cause side effects in some.

Introduction

As a naturalist I am always on the lookout for wild herbs that will add exciting texture and flavor to any meal. When prepared properly their uses are many and varied. The most common uses of wild herbs include jams, jellies, raw or cooked vegetables, fritters, flour, pickling, seasonings, coffee, tea and cold beverages—all are exceptionally fine choices.

Wild herbs can be found virtually everywhere, but exactly where and when should we look for them? Let's begin with where to look. Prime locations for finding wild herbs are country roads and meadows, fields and pastures, the forest floor, and river, lake and stream borders. Those who wish to set out and explore such places will undoubtedly discover a bounty of edible rewards.

When should we harvest wild herbs? Some herb species can be harvested year round such as the common dandelion (*Taraxacum officinale*) or the common cattail (*Typha latifolia*). Others like common plantain (*Plantago major*) can be taken in early spring or common burdock (*Arctium minus*) in the spring and early summer. With time, practice and patience, you will soon learn where and in what season to locate your favorite wild herbs.

What about tools for collecting? All you really need is a good field guide that includes the poisonous plants of your area (to know what plants to avoid), a pocket knife for digging and cutting, and leather gloves for handling plants that can cause dermatitis; such as the stinging nettle (*Urtica dioica*). Harvesting wild herbs is something anyone can do, so before we begin our adventure in locating the feasts of the fields let's talk about some precautions.

It is very important, and I cannot stress this enough, that the plants intended for use should always be identified carefully—if there is any uncertainty about the plants identity—leave it!

Avoid ill-looking plants and those that have been exposed to insecticides and those that grow along heavily traveled roads. Avoid all plants that grow in or near contaminated water sources.

Harvest plants only in their appropriate season and never eat any plant excessively. Never collect rare or endangered plants, and don't trespass onto private lands without permission. Before you prepare wild herbs there are a few guidelines to consider that will help ensure the finest quality and distinct flavors come through.

Plants that are harvested should be prepared as soon as possible to ensure quality and freshness. Always rinse plants thoroughly in cold water before cooking. When cooking herbs make sure that the water is boiling during all cooking stages; never overcook and use only enough water to completely cover the parts used to ensure optimum flavor. Most wild herbs have a distinct bitter taste; salted water, as well as changing the water several times during cooking will usually tame this quality.

When preparing herbal teas with dried materials, use one teaspoon for every six ounces of boiled water. Pour water over the dried material and steep as long as fifteen minutes. Strain the materials from the tea and sweeten with honey or sugar. Some fresh herb materials can be prepared in the same way by using two teaspoons for every six ounces of boiled water. When preparing herbal coffee, use one-half cup of dried ground herb for every four cups of cold water; use a clean coffee pot and never overbrew.

Collecting wild herbs is a lot of fun whether it is to study them or eat them. However, great care must be taken since some wild herbs resemble poisonous ones and the ingestion of some wild herbs may cause side effects. If there is any doubt, consult an expert. Medicinal values have been included since every herb contains a healing property. However, this guide is not a prescriptor and is not to be used as such; consult your physician.

By following these simple precautions and guidelines you should have little trouble discovering, identifying, harvesting and preparing the twenty-eight wild herbs that I have found to be the most versatile and the most flavorful—happy hunting!

Finally, I would like to thank those people who helped make this book possible. Barbara Krieg of the Portland Parks Bureau of Maintenance was a great help in answering all my questions. I also wish to thank my parents for their continuous support and encouragement and making me realize that dreams really do come true. Most of all, I want to thank my husband Christopher for his understanding and never-ending support. And lastly, my thanks goes to the loving memory of my grandmother Emma, who loved the outdoors and cherished her garden, she was an inspiration and a blessing—this book is for her.

Chickweed
Stellaria media

Flowers: white, one-quarter inch wide.
In bloom: February–December.
Life cycle: perennial.
Size: two and one-half feet long.

Leaves are small, ovate, long-stemmed and grow opposite. The stalks are vine-like, twisting and trailing along the ground. Tiny recurved hairs grow from only one side of the stems. Flowers are five-petaled, white and have divided petals that are taller than the sepals. Seeds are reddish brown, round, notched at one end and covered with tiny bumps. Roots develop at the nodes.

Habitat: roadsides, gardens.

Edible parts: leaves, stalk, stems.

Harvest: spring through fall as edible parts become available.

Use: leaves, stalks and stems can be boiled for five minutes in salted water and then served with butter or prepared like creamed spinach. Stalks and stems can be added to cooked vegetables. Raw stalks and stems can be washed, sliced and added to freshen garden and potato salads.

Medicinal value: acne, coughs, circulation, constipation.

©95

9

Field Peppergrass
Lepidium campestre

Flowers: white, minute.
In bloom: May–September.
Life cycle: annual.
Size: six to eighteen inches tall.

Lower leaves form a basal rosette that are ovate; upper leaves are lance-shaped, toothed and clasp the stalk. Flowers are four-petaled, white and are borne in spiked clusters. Seedpods are green when young, turning golden brown when mature. They are flat, oval and notched at one end. Each seedpod contains two seeds.

Habitat: fields, gardens.

Edible parts: leaves, seedpods.

Harvest: spring leaves; summer seedpods.

Use: leaves can be boiled for ten to fifteen minutes and then added to additional cooked greens, cooked vegetables or casseroles. Raw leaves can be added to potato, pasta or garden salads. Leaves are rich in iron and protein, as well as vitamin C. Seedpods can be used in place of pepper to flavor soups, salads, meats, fish and poultry dishes. (For best results, harvest young leaves and green seedpods.)

Medicinal value: scurvy.

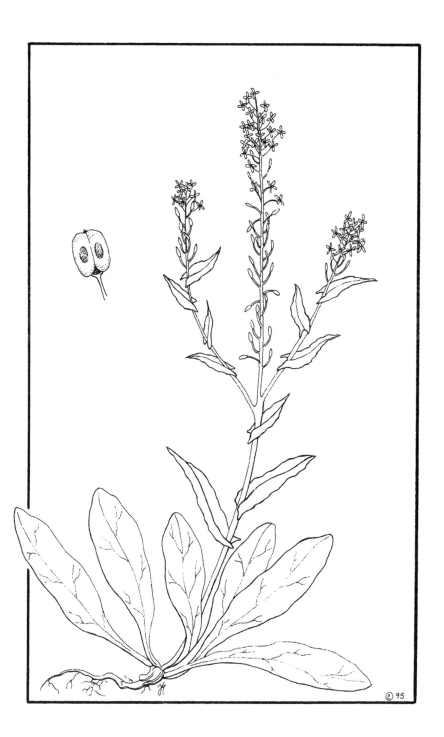

Ox-Eye Daisy
Chrysanthemum leucanthemum

Flowers: white, two and one-half inches wide.
In bloom: May–October.
Life cycle: biennial.
Size: one to three feet tall.

First year plants produce a low basal rosette of spoon-shaped leaves. Second year plants produce the stalk and flowers. Leaves are dark green with irregular lobes and alternate on the stalk. The stalk is slender and often branching. Flowers are a ray of white petals that embrace a yellow disk. Seeds are oblong, gray and have dark ribs.

Habitat: roadsides, fields, pastures, meadows, dry soil.

Edible parts: young leaves, flowers.

Harvest: spring.

Use: raw young leaves can be added to garden salads. However, because of their intense bitterness, use them sparingly. (Young leaves are light green and produce the best flavor.) Flowers should be harvested as soon as they have opened; they can then be used raw or dried for making an herbal tea. Steep the fresh or dried flowers for five to ten minutes in boiled water; strain and then sweeten with honey or sugar.

Medicinal value: colds, coughs.

Queen Anne's Lace
Daucus carota

Flowers: white, one-eighth inch wide.
In bloom: May–September.
Life cycle: biennial.
Size: one to three feet tall.

Leaves are fernlike and alternate on a hairy stalk. Flowers are white and lacy and clustered together forming large, umbel-shaped flower heads. In the center of each flower head is a tiny, burgundy red flower. Dried flower heads resemble birds' nests that hold numerous greenish brown seeds that bear four rows of spines. The root is creamy white, thick and smells like a carrot.

Habitat: roadsides, fields, pastures, vacant lots.

Edible parts: root, seeds.

Harvest: spring root; fall and winter seeds.

Use: the root can be washed, sliced and boiled until tender, about fifteen minutes; it can then be served as a cooked vegetable in place of carrot. The root can be dried in the oven at 180°F until dark brown and dry throughout; it can then be ground and prepared as an herbal coffee. Raw seeds can be steeped ten to fifteen minutes in boiled water, strained and then served as an herbal tea. They may also be used as a spice to season soups, salads and meat dishes.

Medicinal value: coughs, laxative.

Warning! Do not confuse with water hemlock (*Cicuta maculata*) or poison hemlock (*Conium maculatum*) that look similar; both are deadly poisonous.

15

Strawberry
Fragaria virginiana

Flowers: white, three-quarters inch wide.
In bloom: June–August.
Life cycle: perennial.
Size: three to ten inches tall.

Leaves are long stemmed, toothed and compound. The stalks, stems and leaves are hairy. Flowers are five-petaled, white and are borne atop independent stems that do not exceed above the leaves. Berries are miniature replicas of the cultivated strawberry. The main root is thick and shallow with many rootlets attached. Plants spread by producing trailing runners with shallow rootlets that form at the nodes.

Habitat: fields, woods, woodland edges.

Edible parts: leaves, berries, root.

Harvest: throughout the summer as edible parts become available.

Use: leaves must be dried, they can then be steeped ten to fifteen minutes in boiled water, strained and served as an herbal tea. (Raw leaves are poisonous.) Berries can be prepared in the same manner as store-bought varieties; their flavor is superb. The root can be dried in the oven at 175°F until brittle, sliced and then steeped ten to fifteen minutes in boiled water; then strained and served as an herbal tea.

Medicinal value: diarrhea, sore throats, stomachaches.

Warning! Raw leaves are poisonous.

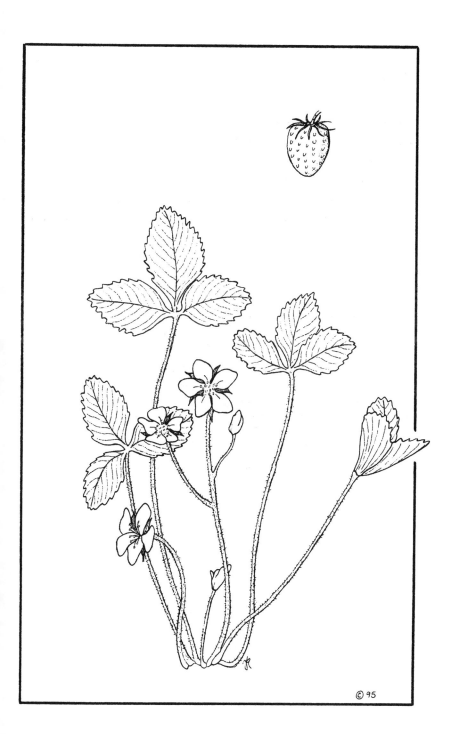

© 95

17

Watercress
Nasturtium officinale

Flowers: white, one-quarter inch wide.
In bloom: March–November.
Life cycle: perennial.
Size: six inches to ten feet long.

Leaves are dark green, glossy, ovate and are divided with as many as nine leaflets. Leaves grow opposite on erect stems that join at the nodes on a creeping stalk. Flowers are four-petaled, white and form loose clusters. Seedpods are slender, semitransparent and erect. Roots develop at the nodes.

Habitat: streams, springs.

Edible parts: leaves, stems.

Harvest: year-round as edible parts become available.

Use: raw leaves and stems can be chopped and used to flavor soups, potatoes, pasta and various salad dishes. (Its flavor is extremely pungent, so use sparingly.) Leaves can also be dried and then sprinkled onto foods for a variety of flavor.

Medicinal value: appetite stimulant, blood purifier.

Warning! Excessive or prolonged use may cause nausea.

© 95

19

White Clover
Trifolium repens

Flowers: white, one inch long.
In bloom: March–October.
Life cycle: perennial.
Size: four to fifteen inches tall.

Leaves are compound on independent stems that stand erect and join at the nodes on a creeping stalk. Leaves have toothed edges and bear yellowish V-shaped bands halfway up the midrib. Flowers are white, often tinged with pink and are pea-like. Each flower clusters together to form independent flower heads that are supported by independent stems that join at the nodes. Seeds are minute and concealed within the dried flower heads. Roots are shallow and trailing and form at the nodes.

Habitat: roadsides, fields, pastures, lawns, gardens.

Edible parts: entire plant.

Harvest: spring through summer as the plant becomes available.

Use: stems, leaves, flowers and roots should be soaked in cold, salted water for twenty to thirty minutes to tenderize, making digestion easier. They can then be added raw to garden, potato or pasta salads. The entire plant may also be boiled for five to ten minutes and then mixed with cooked vegetables or served as cooked greens topped with butter. Flowers can be dried and then steeped in boiled water fifteen minutes, strained and then served as an herbal tea or cooled for an iced tea. Dried flowers can also be ground and blended with whole wheat flour or fried with potatoes for a unique texture and flavor.

Medicinal value: colds, coughs, fevers, gout.

Yarrow
Achillea millefolium

Flowers: white, one-half inch wide.
In bloom: June–September.
Life cycle: perennial.
Size: one to three feet tall.

Leaves are soft and fernlike and alternate on woolly stalks; their scent is strong but pleasant. Flowers are five-petaled, white and embrace a yellow disk. Flowers form independent clusters that deceivingly look like single flowers; these flower heads are umbel-shaped. Seeds are grayish brown and winged.

Habitat: roadsides, fields, meadows, open woods.

Edible parts: leaves.

Harvest: summer.

Use: leaves can be dried and then steeped in boiled water for five minutes; strained and then served as an herbal tea, sweetened with honey or sugar. Juice extracted from the leaves can be taken internally; add one teaspoon of juice to two teaspoons of cold water. Its properties are high in protein and chlorophyll.

Medicinal value: blood purifier, colds, congestion, fevers, stomach cramps.

Warning! Do not mistake for fool's parsley (*Aethusa cynapium*) or poison hemlock (*Conium maculatum*) that are fatal if ingested.

Chicory
Cichorium intybus

Flowers: blue, white or pink, one and one-half inches wide.
In bloom: May–October.
Life cycle: perennial.
Size: two to four feet tall.

Lower leaves form a basal rosette and are deeply lobed. Upper leaves are slightly hairy, linear and are much smaller. The stalk is hairy with branching stems that support blue, white or pink flowers. The flowers are stemless and bloom from the leaf axils; their petals are daisylike with fringed tips. Seeds are golden brown with flattened tops. The root is thick and white.

Habitat: roadsides, fields, vacant lots.

Edible parts: leaves, root.

Harvest: spring leaves; spring and fall root.

Use: leaves can be boiled for ten minutes and served with additional cooked greens or topped with butter. The raw, white, fleshy portion of the underground leaves makes a wonderful addition to garden and potato salads. The root can be baked in the oven at 180°F until dark brown and dry throughout, it can then be ground for making an herbal coffee.

Medicinal value: appetite stimulant, blood purifier, digestion stimulant.

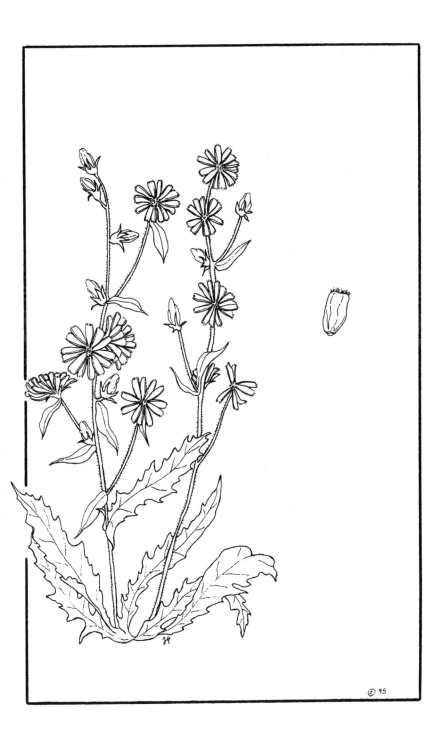

© 95

Musk Mallow
Malva moschata

Flowers: white to pink, one and one-half inches to two inches wide.
In bloom: June–September.
Life cycle: annual.
Size: one to two feet tall.

Leaves are divided, deeply lobed and alternate on hairy stalks and stems. Flowers are five-petaled and are white to pink; their petals are notched. The fruit is green, segmented and enclosed within the calyx. Seeds are tan, rounded and notched.

Habitat: roadsides, fields, cultivated ground.

Edible parts: leaves, fruit, flower buds.

Harvest: spring leaves; summer fruit, flower buds.

Use: leaves can be boiled for fifteen minutes and served as cooked greens topped with butter. Raw, young leaves can be used in place of okra to thicken soups and stews. Leaves may also be dried and then steeped ten to fifteen minutes in boiled water and then strained for an herbal tea, but its medicinal properties are not as effective as other mallows. Flower buds can be pickled and used as capers to flavor sauces and garnishes. Fruits make wonderful additions to garden or fruit salads.

Medicinal value: bronchitis, coughs, stomachaches.

Nodding Wild Onion

Allium cernuum

Flowers: white to pink, one-quarter inch to one and one-half inches wide.
In bloom: July–August.
Life cycle: perennial.
Size: six inches to two feet tall.

Leaves are long and linear (grasslike) and spring upward from the base of a slender smooth stalk. Flowers are white, occasionally pink, with six petals. Flowers and bulbets cluster together nodding toward the ground; aged flower heads and bulbets stand erect. The underground bulb is reddish pink with hair-like roots. The entire plant smells of onion.

Habitat: meadows, prairies, open woods.

Edible parts: leaves, bulb.

Harvest: year-round bulb; spring leaves.

Use: raw leaves can be added to garden salads. Bulbs can be boiled five to ten minutes or steamed until tender. They can then be mixed with additional cooked vegetables or used to flavor soups, stews and meat dishes. They may also be sauteed three minutes in butter and then poured over rice. Raw bulbs can be added to potato, pasta and garden salads or pickled for future use. (Leaves harvested before the flowers develop produce the best flavor.)

Medicinal value: digestion stimulant.

Warning! Those with sensitive stomachs should avoid the onion.

Bull Thistle
Cirsium vulgare

Flowers: pink to purple, one and one-half inches to three inches wide.
In bloom: June–October.
Life cycle: biennial.
Size: two to four feet tall.

Leaves are deeply lobed with sharp spines and alternate on a spiny stalk. Flowers vary from purple to pink and are supported by spiny green bracts that bear yellow tips. Seeds are golden-brown and parachutelike.

Habitat: roadsides, fields, pastures, gardens, fence rows.

Edible parts: leaves, stalk.

Harvest: spring through fall as edible parts become available.

Use: leaves should be cleaned of their spines; they can then be boiled for five to ten minutes and served as cooked greens topped with butter or a vinaigrette. Raw leaves with their spines removed can be added to garden salads. Leaves with their spines removed can be dried and then steeped in boiled water for ten to fifteen minutes, strained and then served as an herbal tea, sweetened with honey or sugar. The stalk can be peeled to remove the spines and then sliced and boiled ten to fifteen minutes and served as a cooked vegetable over rice.

Medicinal value: diarrhea, skin disorders.

Warning! Wear thick leather gloves when harvesting to avoid skin contact with the plant's sharp spines.

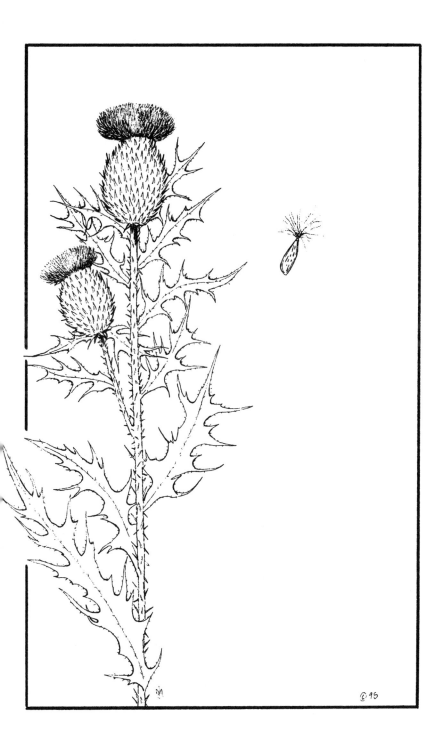

Burdock
Arctium minus

Flowers: pink to purple, one-half inch to one inch wide.
In bloom: July–October.
Life cycle: biennial.
Size: two to five feet tall.

First-year plants produce a low rosette of large woolly leaves. Second-year plants produce the stalk and flowers. Leaves grow opposite and are heart-shaped with reddish stems. The stalk is tinged pink and is hairless. Flowers are pink to purple and are borne atop green, dime-sized burs. Flower heads are supported by short independent stems. Burs have spiny, recurved hooks that contain numerous oblong-shaped, ridged seeds. The root is long and thick.

Habitat: roadsides, fields, woodland edges.

Edible parts: leaves, leafstalks, root.

Harvest: spring leaves, leafstalks; early summer root.

Use: leaves can be boiled for fifteen minutes in two changes of water, they can then be served with butter or prepared like creamed spinach. Raw or dried leaves can be steeped five to ten minutes in boiled water and then strained for an herbal tea or cooled for an herbal iced tea. Raw leafstalks can be peeled and added to garden salads or they can be boiled ten to fifteen minutes and prepared like creamed asparagus. The root can be peeled, sliced and boiled for ten to fifteen minutes and then served with butter or added to mixed vegetables. It may also be sauteed with onions in butter or vegetable oil for five minutes and then served over rice. Raw root peeled and sliced can be added to salads in place of celery. The root may also be dried in the oven at 175°F until brittle. Sliced, it can then be steeped ten to fifteen minutes in boiled water, strained and then served as a hot or cold tea, sweetened with honey or sugar.

Medicinal value: blood purifier, fevers, indigestion, lowering blood sugar levels.

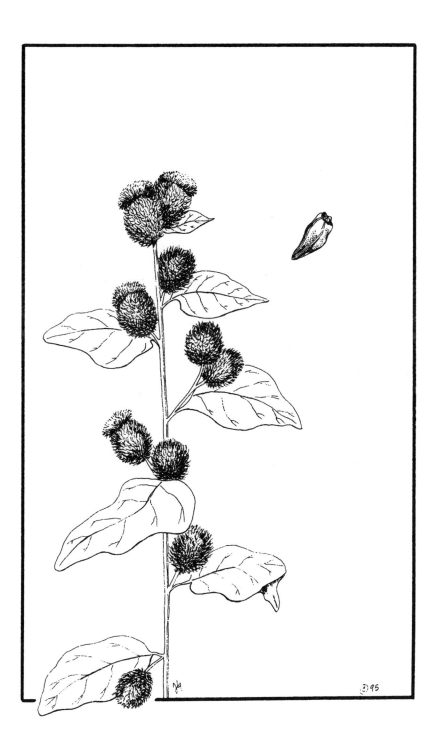

Fireweed
Epilobium angustifolium

Flowers: pink to red, three-quarters inch to one and one-half inches wide.
In bloom: June–September.
Life cycle: perennial.
Size: three to seven feet tall.

Leaves are toothless, short-stemmed and alternate on a hairless stalk. Flowers are four-petaled and vary in hue from hot pink to red; they are borne in spikes. Seedpods are three inches long, reddish brown and slender; they angle upward. Each seedpod is filled with rows of downy seeds that quickly become airborne when the pods dry and split open.

Habitat: roadsides, open woods, burned land.

Edible parts: shoots, leaves.

Harvest: spring shoots; summer leaves.

Use: shoots can be boiled ten to fifteen minutes and then served like asparagus or they can be added to cooked vegetables and served over rice. Young leaves can be boiled five to ten minutes and then served with a cream sauce. Mature leaves can be dried and steeped for fifteen minutes in boiled water, strained and then served as an herbal tea, sweetened with honey or sugar.

Medicinal value: cramps, diarrhea.

Milkweed
Asclepias syriaca

Flowers: pink to red, one-half inch wide.
In bloom: June–August.
Life cycle: perennial.
Size: two to six feet tall.

Leaves are broadly ovate and grow opposite. Above, the leaves are dark green, below they are grayish green and woolly. Flowers are star-shaped; their color may vary from pale pink to red. Flowers cluster together to form single flower heads. Seedpods are grayish green when young and are covered with tiny knobs. Mature seedpods split open to release numerous, downy white, parachutelike seeds. The shoots are woolly. Stalk, leaves and shoots exude a poisonous milky liquid when torn or broken.

Habitat: roadsides, fields, fence rows, pastures, dry soil.

Edible parts: leaves, shoots, flower buds, flowers, young seedpods.

Harvest: spring leaves, shoots; summer flower buds, flowers, young seedpods.

Use: leaves, shoots, flower buds and young seedpods must be boiled fifteen minutes in several changes of water to expel their toxins. The leaves and flower buds can then be added to additional cooked greens. Shoots can be prepared like asparagus and the seedpods can be added to cooked vegetables. The flowers should be boiled for about sixty seconds, they can then be batter dipped and fried or blended with cooked vegetables.

Medicinal value: asthma, kidney stones, stomach ailments, water retention.

Warning! Do not confuse with butterfly weed (*Asclepias tuberosa*) which does not exude a milky liquid, or any of the various dogbanes (*Apocynum medium* or *Apocynum cannabinum*) which have smooth stalks; these plants can be fatal if ingested. (With excessive or prolonged use the common milkweed is poisonous.)

Red Clover
Trifolium pratense

Flowers: pink to red, one inch wide.
In bloom: April–October.
Life cycle: perennial.
Size: six to sixteen inches tall.

Leaves are long-stemmed and compound. Each leaflet bears a prominent cream-colored band in the shape of a V across the midrib. Flowers are reddish pink and pea-like, and form one inch flower heads whose stems join at the leaf axils. Seeds are minute and concealed within the dried flower heads.

Habitat: roadsides, fields, pastures, lawns.

Edible parts: leaves, flowers, seeds.

Harvest: spring through summer as edible parts become available.

Use: raw leaves and flowers should be boiled for ten minutes or soaked in salted water for several hours to make them digestible. They can then be added to additional cooked greens or vegetable dishes. Flowers and their seeds can be dried and then steeped ten to fifteen minutes in boiled water and then strained for an herbal tea or an iced tea. Dried flowers and their seeds can be ground and then added to whole wheat flour. Dried flowers can also be fried with potatoes for additional texture and flavor.

Medicinal value: asthma, bronchitis, blood purifier, coughs.

Dandelion
Taraxacum officinale

Flowers: yellow, two inches wide.
In bloom: year-round.
Life cycle: perennial.
Size: six inches to two feet tall.

Leaves are irregular in size and shape with deeply pointed lobes and form a tight rosette. Stalks are erect, slender, hollow and unbranching. Each stalk produces a single canary yellow flower. Seedheads are golf ball-sized and can produce as many as 139 downy white, parachutelike seeds. The root is rust orange and can exceed twelve inches in length.

Habitat: roadsides, fields, pastures, lawns, gardens, vacant lots.

Edible parts: leaves, flower buds, flowers, stalks, root.

Harvest: year-round as edible parts become available.

Use: leaves can be boiled or steamed five to ten minutes and then served with butter or a cream sauce. Raw leaves make a wonderful addition to garden salads or served on their own with a vinaigrette. Flower buds can be boiled or steamed five to ten minutes and then topped with butter or added to additional cooked vegetables. They are excellent when sauteed in butter for several minutes and served over rice. Batter dipped and fried their flavor resembles that of broccoli. Flower buds may also be pickled and stored for future use. Raw stalks can be sliced and added to garden salads. The root can be wiped clean and then baked in the oven at 180°F until dried and brittle, it can then be ground for making an herbal coffee. For an herbal tea or iced tea, chop the dried root and steep in boiled water for ten minutes, strain and serve, sweeten with honey or sugar.

Medicinal value: anemia, appetite stimulant, bronchitis, fevers, indigestion.

Evening Primrose
Oenothera biennis

Flowers: yellow, one to two inches wide.
In bloom: June–October.
Life cycle: biennial.
Size: one to five feet tall.

First-year plants produce a low basal rosette. Second-year plants are hairy and produce a tall, leafy, slender stalk with branching reddish stems that support yellow flowers. Leaves are lance-shaped and grow opposite. Flowers are four-petaled and produce a faint scent of lemon; their sepals are shorter than the stigma which bears a distinct X shape. Seeds are enclosed in pods that resemble wooden flowers with four curled petals. Seeds are irregular in shape and resemble coffee grinds. The root is thick and reddish pink.

Habitat: roadsides, fields, thickets, dry soil, sandy soil.

Edible parts: young leaves, root.

Harvest: spring young leaves; fall root.

Use: young leaves should be peeled and then boiled twenty minutes in two changes of water, they can then be topped with butter or mixed with additional cooked greens. Raw young leaves produce a mild peppery flavor. Raw, peeled young leaves make an interesting addition to garden salads. The root should be peeled, sliced and then boiled for thirty minutes in three changes of water, its flavor is like that of pepper. It makes a wonderful cooked vegetable served with butter or when mixed with cooked corn and carrot. If left to cool it can be added to garden or potato salads.

Medicinal value: bronchitis, constipation, headaches.

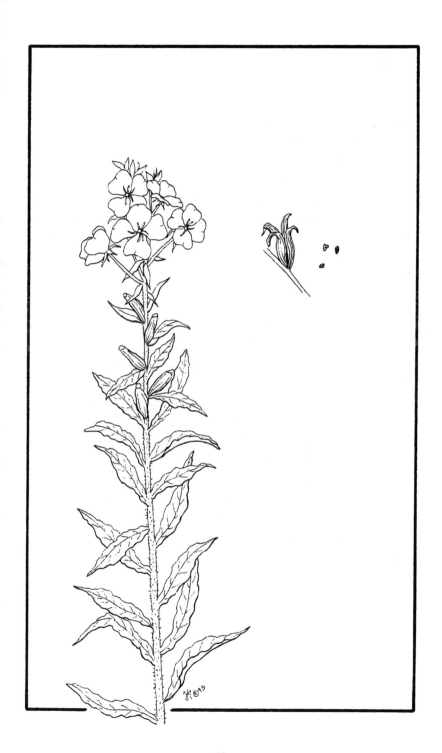

43

Mullein
Verbascum thapsus

Flowers: yellow, one-quarter inch to one inch wide.
In bloom: May–September.
Life cycle: biennial.
Size: one to six feet tall.

First-year plants produce a low rosette of large woolly leaves. Second-year plants produce a tall, furry, leafy, flowering stalk. Leaves are silvery green, woolly and can reach a length of fifteen inches. Flowers are five-petaled, yellow and bloom at random. Seeds are oblong with ridged wavy lines.

Habitat: roadsides, fields, vacant lots.

Edible parts: leaves, flowers.

Harvest: summer leaves, flowers.

Use: leaves and flowers can be dried and then steeped ten to fifteen minutes in boiled water, strained twice and then served as an herbal tea or cooled for an iced tea. Its flavor is excellent. (After the tea has been prepared it is important that it be strained several times through a muslin cloth before drinking to eliminate the hairs that can cause severe itching in the mouth.)

Medicinal value: bronchitis, colds, coughs, sore throats, sinus.

Warning! First-year rosette resembles first-year foxglove (*Digitalis purpurea*) that is fatal if ingested. Handling may cause dermatitis. Will cause severe itching in the mouth if not prepared properly.

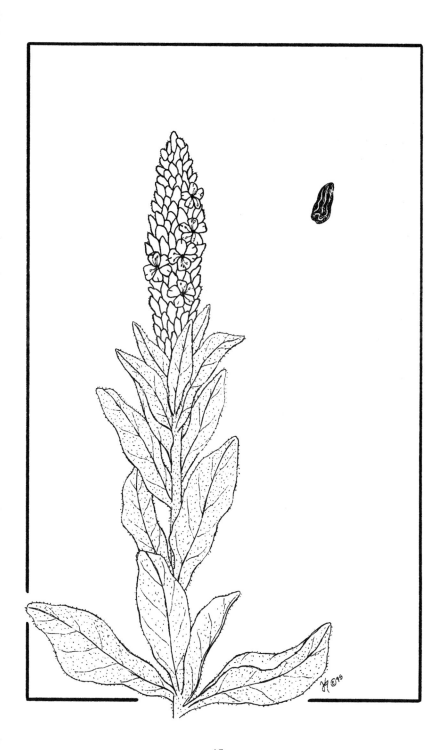

Purslane
Portulaca oleracea

Flowers: yellow, one-half inch wide.
In bloom: May–October.
Life cycle: annual.
Size: one to two feet tall.

Leaves are light green, wedge-shaped and fleshy. Stalk and stems are reddish pink and smooth. Flowers are five-petaled and yellow; they are stemless and bloom from the center of leafy rosettes on independent stems. Seeds are black, oval and coarse.

Habitat: fields, lawns, gardens, sandy soil.

Edible parts: entire plant.

Harvest: summer.

Use: the entire plant can be simmered for ten minutes in salted water and then served with butter or chopped and added to cooked vegetables. Raw it may be used for pickling or it can be chopped and added to garden salads. It also makes a wonderful addition when added to soups and stews. Leaves and stems make an interesting addition when mixed with green beans and vinegar or when they are added to pasta or potato salad. Seeds can be dried and ground for making flour.

Medicinal value: headaches, stomachaches.

Scotch Broom

Cytisus scoparius

Flowers: yellow, one inch wide.
In bloom: May–June.
Life cycle: evergreen shrub.
Size: three to five feet tall.

Leaves are dark green, compound and are one-half inch long.
Stems are also dark green, grooved, stiff and very strong.
Flower buds are small and hairless. Flowers are mustard yellow
with pea-like petals and bloom profusely nearly concealing the
leaves and stems. Seedpods are green when young, turning
dark brown with age. Their size is three-quarters inch to two
inches long, they are flat, hairy and hang downward. The seeds
within the pods are bean-shaped and notched at one end.

Habitat: roadsides, vacant lots, dry soil.

Edible parts: flower buds, young seedpods, seeds.

Harvest: spring seedpods; summer flower buds, seeds.

Use: young seedpods and flower buds must be soaked thoroughly for at least twelve hours in a strong solution of salt and
vinegar to rid them of their toxins. They can then be rinsed
thoroughly in cold water and used for pickling. Seeds can be
roasted in the oven at 180°F until dark brown and then ground
and prepared as an herbal coffee.

Medicinal value: constipation, high blood pressure, toothaches.

Warning! Raw materials of this herb contain toxins that can
cause fatal poisoning. Prepared materials–if consumed excessively or in large doses–can also cause fatal poisoning.

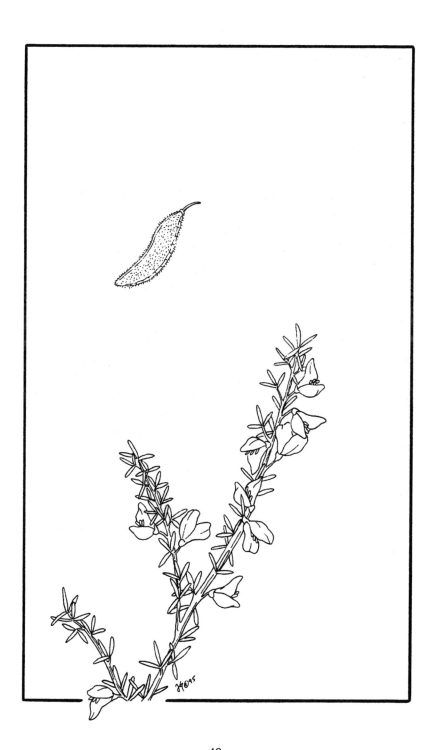

Wild Lettuce
Lactuca canadensis

Flowers: yellow, one-half inch to three-quarters inch wide.
In bloom: June–October.
Life cycle: biennial.
Size: one to ten feet tall.

Leaves are lance-shaped, toothed and deeply lobed. The stalk is smooth and covered by a filmy, off-white skin. Flowers are yellow and resemble those of the dandelion and form loose, flowering clusters on branching stems. Seeds are downy white parachutes with slender stems and a ridged beak.

Habitat: roadsides, fields, meadows, open woods.

Edible parts: leaves, flower buds.

Harvest: spring leaves; summer flower buds.

Use: leaves can be boiled ten to fifteen minutes in two changes of water and then topped with butter or added to cooked greens. Raw leaves can be added to garden salads or they can be mixed with dandelion greens and topped with a vinaigrette. Raw flower buds are bitter, but when added to garden salads or casseroles they present a unique flavor.

Medicinal value: headaches, sedative.

Warning! This herb may cause internal poisoning to some. Excessive use may also cause internal poisoning. Handling of this herb may cause dermatitis. Consult your physician before using this herb.

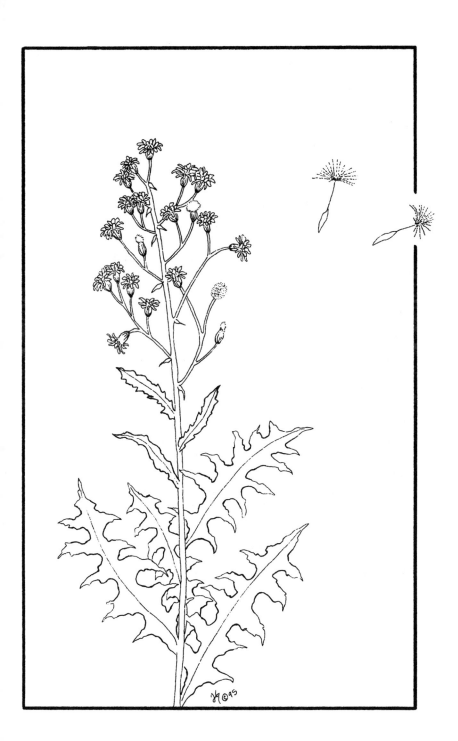

Winter Cress
Barbarea vulgaris

Flowers: yellow, one-half inch wide.
In bloom: April–August.
Life cycle: perennial.
Size: one to two feet tall.

Leaves are dark green and glossy. Lower leaves form a basal rosette; they are rounded and deeply lobed. Upper leaves are toothed, clasping and alternate. Flowers are four-petaled, yellow and cluster together. Flower heads are supported by individual branching stems that join at the leaf axils. Seeds are greenish brown and are divided at one end.

Habitat: roadsides, fields, moist ground.

Edible parts: leaves, flower buds.

Harvest: winter and spring leaves; spring flower buds.

Use: leaves can be boiled for ten minutes in three changes of water. They are best when mixed with other cooked greens since they are extremely bitter. Young leaves of the rosette have a much milder flavor and are excellent when added to garden salads. Boiled for five minutes they can be served like spinach. Flower buds should be boiled for five minutes in two changes of water to rid them of their bitterness, they can then be topped with butter and served with rice or mixed with cooked corn.

Medicinal value: appetite stimulant, coughs.

Warning! May cause kidney damage. Consult your physician before using this herb.

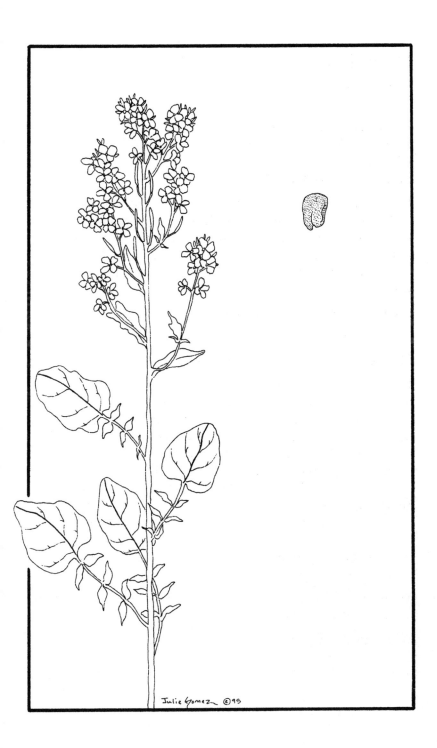

Julie Gomez ©95

Curly Dock
Rumex crispus

Flowers: green, minute.
In bloom: May–September.
Life cycle: perennial.
Size: one to five feet tall.

Lower leaves of the rosette are long, dark green, very coarse and have curled margins. Upper leaves are short and slender with curled margins. Leaves alternate on a greenish brown stalk. Flowers are green, tinged with red and form loose clustered spikes that join at the leaf axils. Seeds are reddish brown with heart-shaped wings.

Habitat: roadsides, fields, pastures, meadows.

Edible parts: leaves, seeds, root.

Harvest: spring leaves; summer root; fall seeds.

Use: leaves can be boiled in two changes of lightly salted water for fifteen minutes and then served like spinach. Cooked greens are best when mixed with cooked dandelion greens and then topped with butter. Young leaves can be used raw to flavor garden salads and should be gathered before the stalks appear, otherwise they are much too bitter. The root can be dried in an oven at 180°F until brittle, it can then be sliced and boiled for ten minutes and then steeped in boiled water for an additional ten minutes; the liquid can then be strained through a muslin cloth and served as an herbal tea or cooled for an iced tea. Seeds can be prepared into flour after their husks have been removed.

Medicinal value: anemia, blood purifier, coughs, constipation, headaches, skin disorders.

Warning! Excessive use may cause diarrhea, nausea.

Lamb's Quarters
Chenopodium album

Flowers: green, minute.
In bloom: June–October.
Life cycle: annual.
Size: one to three feet tall.

Leaves are dark green above and pale below and alternate on the stalk. Lower leaves are wedge-shaped and broadly toothed while the upper leaves are linear and toothless. Young leaves have a white, mealy coating. Stalk and stems are hairy and greenish pink. Flowers are five-petaled, green and form loose spiked clusters that join at the leaf axils. Seeds are dark brown with rounded edges and are enclosed within the calyx.

Habitat: roadsides, fields.

Edible parts: leaves, seeds.

Harvest: summer leaves; fall seeds.

Use: leaves can be boiled or steamed for ten minutes and then mixed with cooked vegetables or they can be served as cooked greens and topped with butter, vinegar or a cream sauce. Raw, young leaf tips are excellent and can be added to garden salads. Leaves can be used in place of lettuce for sandwiches or chopped and added to chicken, pasta or potato salads. Leaves can be dried and steeped fifteen minutes; the liquid can then be strained and served as an herbal tea. When cooled it makes a refreshing iced tea. Seeds can be sifted to remove their husks and then crushed or used whole to be sprinkled on breads prior to baking. They may also be ground and then mixed with whole wheat flour.

Medicinal value: diarrhea, iron deficiency, stomachaches.

Plantain
Plantago major

Flowers: green, minute.
In bloom: June–October.
Life cycle: perennial.
Size: four to eighteen inches tall.

Leaves form a loose rosette; they are large, ovate, ribbed and slightly hairy. The slender stalks produce independent flower spikes that bear tiny, green, compact flowers. Seeds are minute and are irregular in both size and shape.

Habitat: roadsides, lawns, gardens.

Edible parts: leaves.

Harvest: spring.

Use: leaves can be boiled for fifteen minutes and then topped with butter or a cream sauce and seasoned with pepper. They may also be chopped and mixed with additional cooked greens or vegetables. Raw leaves can be chopped and added to garden salads. Leaves may also be dried and then steeped fifteen minutes in boiled water and then strained for an herbal tea or cooled for an iced tea, sweetened with honey or sugar.

Medicinal value: blood purifier, bronchitis, coughs, diarrhea.

Julie Gomez ©95

Stinging Nettle
Urtica dioica

Flowers: green, one-eighth inch wide.
In bloom: May–October.
Life cycle: perennial.
Size: two to seven feet tall.

Leaves are ovate with a heart-shaped base. The leaves are dark green, hairy, toothed and grow opposite. The stalk is hairy and often branching. Flowers are green and without petals; they hang in delicate drooping chains from the leaf axils. Seeds are green and minute and are concealed within the calyx.

Habitat: roadsides, woods.

Edible parts: shoots, leaves.

Harvest: spring shoots; summer leaves.

Use: leaves and shoots can be boiled for fifteen minutes and then topped with butter and a squeeze of lemon. They may also be added to soups, stews and casseroles. Raw leaves and shoots can be boiled for three to five minutes and then the liquid strained through a muslin cloth and served as an herbal tea. Dried leaves can be steeped fifteen minutes in boiled water, strained and then served as an herbal tea or cooled for an iced tea, sweetened with honey or sugar.

Medicinal value: anemia, asthma, blood purifier, diarrhea, headaches, ulcers.

Warning! The entire plant consists of stinging hairs that will cause severe dermatitis–wear gloves when harvesting. (Cooking and drying of the plant dissolves its stinging properties.)

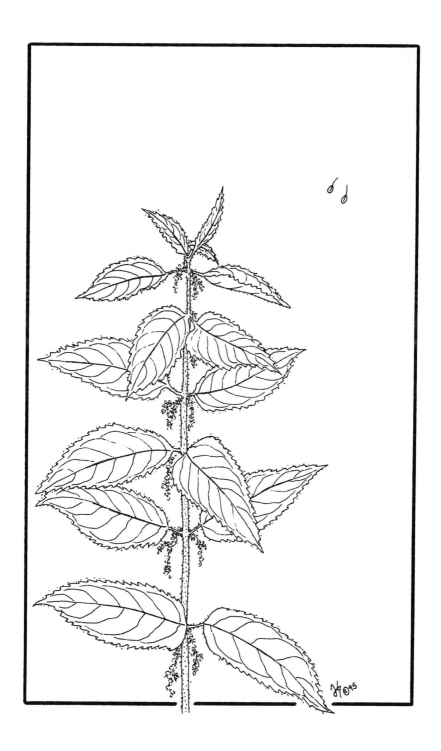

Cattail
Typha latifolia

Flowers: brown, minute.
In bloom: May–July.
Life cycle: perennial.
Size: three to nine feet tall.

Leaves are long and swordlike rising upward from a single stalk. The large cigar-shaped flower head produces brown (female) flowers. Above the flower head is a long slender spike which produces green (male) flowers that later turn a golden-brown when full of pollen. Winter flower heads produce thousands of grayish, billowy seeds.

Habitat: roadsides, marshes, lakes, ponds.

Edible parts: sprouts and their core, shoots, stalk, immature flower head, pollen, rootstock core.

Harvest: spring sprouts, shoots, stalk, immature flower head; summer pollen; fall and winter rootstock.

Use: sprouts can be peeled and boiled for ten minutes or steamed for fifteen minutes, they can then be served as a vegetable or used for pickling. The core can be boiled until firm yet tender and then served like a potato. Cooled and sliced it makes a wonderful addition to potato salads. Shoots and stalks can be peeled and boiled fifteen minutes and then mixed with green beans or served like asparagus. Raw shoots and stalks can be peeled and added to garden and potato salads. Immature flower heads can be boiled until tender and then eaten like corn on the cob. Pollen can be sifted and then blended with equal amounts of whole wheat flour; the mixture can be used in place of pure flour. The rootstock can be peeled to expose the white inner core and then crushed in a pan of cold water. Let the starch settle and remove the floating fibers; repeat this process until there is only pure cattail flour remaining. The flour can be used immediately or dried for storage.

Medicinal value: diarrhea.

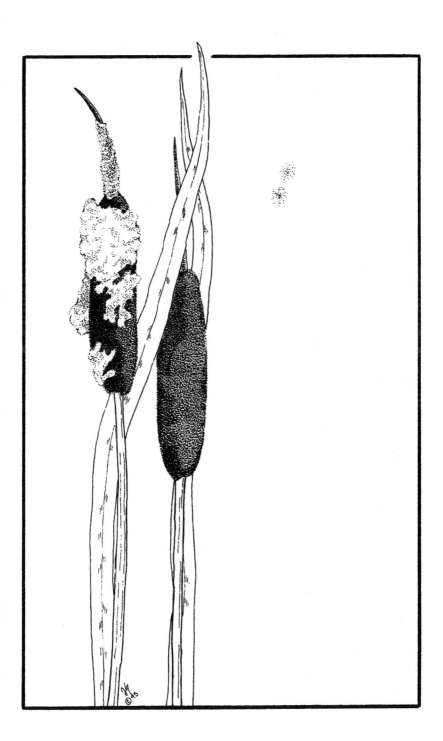

MORE GREAT HANCOCK HOUSE TITLES

Alpine Wildflowers
J. E. (Ted) Underhill
ISBN 0-88839-975-8

Birds of North America
David Hancock
ISBN 0-88839-220-6

**Coastal Lowland
Wildflowers**
J. E. (Ted) Underhill
ISBN 0-88839-973-1

Desert Wildflowers
Mabel Crittenden
ISBN 0-88839-282-6

Eastern Mushrooms
E. Barrie Kavasch
ISBN 0-88839-091-2

Edible Seashore
Rick Harbo
ISBN 0-88839-199-4

**Introducing Eastern
Wildflowers**
E. Barrie Kavasch
ISBN 0-88839-092-0

Indian Herbs
Dr. Raymond Stark
ISBN 0-88839-077-7

NW Native Harvest
Carol Batdorf
ISBN 0-88839-245-1

Orchids of N. America
Dr. William Petrie
ISBN 0-88839-089-0

Pacific Wilderness
Hancock, Hancock & Sterling
ISBN 0-919654-08-8

Rocks & Minerals NW
Learning & Learning
ISBN 0-88839-053-X

Roadside Wildflowers NW
J. E. (Ted) Underhill
ISBN 0-88839-108-0

Sagebrush Wildflowers
J. E. (Ted) Underhill
ISBN 0-88839-171-4

Seashells of the NE Coast
Gordon & Weeks
ISBN 0-88839-086-6

Tidepool & Reef
Rick M. Harbo
ISBN 0-88839-039-4

**Upland Field & Forest
Wildflowers**
J. E. (Ted) Underhill
ISBN 0-88839-174-9

Western Mushrooms
J. E. (Ted) Underhill
ISBN 0-88839-031-9

Western Seashore
Rick Harbo
ISBN 0-88839-201-X

**Wild Berries of the
Northwest**
J. E. (Ted) Underhill
ISBN 0-88839-027-0

Wild Harvest
Terry Domico
ISBN 0-88839-022-X

Wildlife of the Rockies
Hancock & Hall
ISBN 0-919654-33-9